Mastering Mini-Split Systems

Operation And Service Guide For Inverter Mini-Splits

Copyright@2024

Torian Cresswell

Table Of Content

Preface 4

    Introduction To The Book 4

    Who This Book Is For 7

    What The Reader Can Expect To Learn 10

Chapter 1: Introduction To Inverter Mini-Split Systems 14

    1. Overview Of Mini-Split Systems 14

    2. Benefits Of Using Inverter Technology In Mini-Splits 18

    3. Comparison Between Inverter Mini-Splits And Traditional HVAC Systems 22

Chapter 2: Basic Components And Functionality 27

    1. Description Of Key Components (Outdoor Unit, Indoor Unit, Remote Control, Etc.) 27

    2. How Inverter Technology Works 32

    3. Understanding The Refrigeration Cycle In Mini-Splits 37

Chapter 3: Installation Guidelines........................42

   1. Pre-Installation Considerations (Sizing, Location, Etc.)..42

   2. Step-By-Step Installation Process.................47

   3. Common Mistakes To Avoid During Installation...52

Chapter 4: Operating Inverter Mini-Splits...........57

   1. Detailed Guide On Operating The System ...57

   2. Setting Up And Understanding The Control Panel And Remote Operations..........................62

   3. Tips For Optimal Efficiency And Comfort...67

Chapter 5: Maintenance Essentials 72

   1. Routine Maintenance Schedule....................72

   2. Cleaning And Replacing Filters...................77

   3. Checking And Maintaining Refrigerant Levels 82

Chapter 6: Troubleshooting Common Issues.......87

   1. Diagnosing Common Problems 87

Chapter 7: Advanced Service Procedures 98

   1. Technical Service Procedures 98

   2. Tools And Equipment Needed For Advanced Servicing 103

   3. Safety Precautions And Best Practices 108

Chapter 8: Case Studies 113

   1. Real-World Examples Of Troubleshooting And Repair 113

   2. Discussion On Maintaining System Longevity And Performance 119

Chapter 9: The Future Of Inverter Mini-Splits .. 124

   1. Innovations In Inverter Technology And Mini-Split Systems 124

   2. Environmental Impact And Energy Efficiency Trends 129

Appendix 134

   Glossary Of Terms 134

   Frequently Asked Questions (FAQs) 138

Preface

Introduction To The Book

Welcome to "Mastering Mini-Split Systems: Operation and Service Guide for Inverter Mini-Splits." This book is designed to serve as a comprehensive resource for anyone interested in understanding, installing, and maintaining inverter mini-split systems. Whether you are a seasoned HVAC technician, a homeowner looking to make informed decisions about heating and cooling solutions, or a student of HVAC technology, this guide aims to provide valuable insights and practical advice.

Inverter mini-splits represent a significant advancement in climate control technology, offering superior energy efficiency, precise temperature control, and a flexible, less invasive installation process compared to traditional systems. As climate control solutions move towards more sustainable and energy-efficient technologies, understanding how to work with these systems becomes increasingly important.

This book will take you through the various aspects of inverter mini-split systems, from the basic principles of operation and the components involved, to detailed guidelines on installation, troubleshooting, and maintenance. With clear instructions, diagrams, and step-by-step processes, this guide is crafted to make the complexities of inverter technology accessible and easy to understand.

In the following chapters, you will find a thorough exploration of all the essential topics, accompanied by practical tips and professional insights. This will equip you with the knowledge and skills needed to effectively manage and service these modern HVAC systems, ensuring optimal performance and longevity.

Who This Book Is For

"Mastering Mini-Split Systems: Operation and Service Guide for Inverter Mini-Splits" is designed to be an essential resource for a wide range of readers who are interested in the technical and practical aspects of inverter mini-split systems. This includes:

- **HVAC Professionals and Technicians**: For those already working in the field of heating, ventilation, and air conditioning, this book serves as a detailed reference guide for the installation, troubleshooting, and maintenance of inverter mini-splits. It provides advanced insights and technical details that can help professionals enhance their service offerings and expertise.

- **Students and Apprentices in HVAC**: Aspiring HVAC technicians and students enrolled in technical schools will find this book a valuable educational tool. It covers fundamental concepts and advanced procedures in a structured manner, helping to bridge the gap between classroom learning and real-world application.

- **Homeowners and Residential Property Owners**: For individuals who own or are considering installing a mini-split system in their homes, this guide offers a clear understanding of how these systems work, what is involved in maintaining them, and how to troubleshoot common issues, empowering homeowners to make informed decisions about their HVAC systems.

- **Facility Managers and Commercial Property Owners:** Managers responsible for the maintenance and operation of HVAC systems in commercial settings will find practical advice and strategies for managing the efficient operation and maintenance of mini-split systems across larger properties.

- **DIY Enthusiasts**: Even those with a keen interest in DIY home improvement projects can benefit from the clear explanations and detailed installation and maintenance instructions provided in this book.

By providing thorough and accessible information, this book aims to demystify the technology behind inverter mini-splits and equip each of these groups with the knowledge they need to effectively use, maintain, and troubleshoot these systems. Whether you are looking to deepen your professional knowledge or simply aiming to better understand the systems in your home or property, this guide is tailored to help you succeed.

What The Reader Can Expect To Learn

This book is structured to offer readers a comprehensive understanding of inverter mini-split systems, covering a wide range of topics essential for mastering these advanced HVAC systems. Here's what you can expect to learn:

- **Fundamentals of Inverter Technology**: Gain a solid understanding of how inverter technology works and why it offers significant advantages over traditional HVAC systems, including enhanced efficiency and better environmental performance.

- **Components and Design**: Learn about the various components that make up a mini-split system, such as the indoor and outdoor units, remote controls, and inverter-driven compressors. This section will also cover how these components work together to provide effective heating and cooling.

- **Installation Procedures**: Detailed guidance on the proper installation techniques, including choosing the right location, sizing considerations, and step-by-step instructions to ensure the system operates efficiently and effectively.

- **Operational Guidelines**: Insights into the operation of mini-split systems, including how to use and interpret the settings on remote controls and control panels, and tips for optimizing system performance to save energy and extend the lifespan of the unit.

- **Maintenance and Servicing**: Instructions on how to perform routine maintenance, such as cleaning filters and checking refrigerant levels, as well as more advanced servicing tasks to keep the system running smoothly.

- **Troubleshooting Common Problems**: Practical troubleshooting steps for common issues that might arise, such as the system not cooling or heating adequately, unusual noises, or refrigerant leaks, providing tools and techniques to diagnose and resolve problems efficiently.

- **Advanced Repair Techniques**: For more technically inclined readers, this book will delve into sophisticated repair and replacement procedures, including dealing with electronic controls and compressor issues, ensuring safety and compliance with industry standards.

- **Case Studies and Real-World Examples**: To help contextualize the information, the book includes case studies and examples of real-world troubleshooting and repair scenarios, offering insights into how professionals handle complex issues.

- **Future Trends and Innovations**: An exploration of the future direction of inverter technology and mini-split systems, discussing upcoming innovations, potential improvements in energy efficiency, and the evolving regulatory landscape affecting HVAC systems.

By the end of this book, readers will not only have a thorough understanding of inverter mini-split systems but also practical skills and knowledge to install, maintain, troubleshoot, and repair these systems effectively.

Chapter 1: Introduction To Inverter Mini-Split Systems

1. Overview Of Mini-Split Systems

Mini-split systems, also known as ductless systems, have revolutionized the way we think about heating and cooling residential and commercial spaces. They provide a versatile and efficient solution for climate control, especially in areas where traditional ducted systems are impractical or too costly to install. This section provides an overview of mini-split systems, outlining their key features and the various configurations available.

1. Definition and Key Features:

  - **Ductless Nature**: Unlike traditional HVAC systems that rely on extensive ductwork, mini-splits consist of an outdoor compressor/condenser unit and one or more indoor air-handling units. These components are connected by a conduit that houses the power cable, refrigerant tubing, suction tubing, and a condensate drain.

- **Flexibility and Scalability**: Mini-split systems offer the flexibility to heat or cool specific zones in a building. Each indoor unit operates independently, which means different areas can be set to different temperatures, leading to increased comfort and efficiency.

2. **Types of Mini-Split Systems**:

- **Single-Zone Systems**: Consist of one outdoor unit connected to one indoor unit, ideal for single-room applications.

- **Multi-Zone Systems**: Feature one outdoor unit that can be connected to multiple indoor units, each controllable independently, suitable for entire homes or multiple room setups.

3. Advantages of Mini-Split Systems:

- **Energy Efficiency**: Mini-splits are generally more energy-efficient than traditional systems due to their ductless design, which minimizes energy losses typically associated with ductwork.

- **Installation Ease**: With no need for ductwork, installation is less invasive, quicker, and typically less expensive in buildings where ductwork is not present.

- **Quiet Operation**: Mini-splits are known for their quiet operation, making them suitable for environments where noise is a concern, such as bedrooms and offices.

- **Design Flexibility**: Indoor units come in various styles, including wall-mounted, ceiling-mounted, and floor-standing models, providing flexibility to match different interior designs and space requirements.

4. Applications:

- **Residential**: Ideal for new additions, converted attics, garages, or any area where ductwork is not feasible.

- **Commercial**: Commonly used in schools, offices, shops, and other commercial buildings for localized cooling and heating without the need for a centralized ducted system.

- **Retrofitting**: An excellent option for retrofitting older buildings with non-ducted heating systems, such as hydronic, radiant panels, or space heaters.

Understanding the basic configuration and advantages of mini-split systems sets the stage for diving deeper into the specifics of inverter technology, which enhances the efficiency and functionality of these systems even further. T

2. Benefits Of Using Inverter Technology In Mini-Splits

Inverter technology in mini-split systems represents a significant advancement in heating and cooling efficiencies and performance. This technology allows the compressor to operate at variable speeds, adjusting its output to match the specific heating or cooling demands of the space at any given time. Here are the key benefits of using inverter technology in mini-split systems:

1. Energy Efficiency:

   - **Variable Speed Operation**: Unlike traditional compressors, which operate on an all-or-nothing principle, inverter compressors can speed up or slow down based on the needs of the room. This means they consume only the amount of energy needed at any particular moment, leading to considerable energy savings.

- **Reduced Energy Consumption**: By avoiding the energy-intensive on-off cycling typical in traditional systems, inverter systems use less energy overall, which can translate to lower utility bills.

2. Improved Comfort:

  - **Precise Temperature Control**: Inverter mini-splits adjust the compressor's speed to maintain a more consistent temperature within the space. This precision avoids the temperature fluctuations associated with traditional HVAC systems, enhancing overall comfort.

  - **Faster Heating and Cooling**: The ability to operate at higher speeds when needed allows inverter systems to reach the desired temperature setting more quickly than non-inverter models.

3. Extended Lifespan of Equipment:

  - **Less Wear and Tear**: Because the inverter compressor ramps up and down gradually instead of turning on and off abruptly, there is less wear and tear on the system's components. This can extend the overall lifespan of the system.

   - **Operational Efficiency**: Continuous operation at low power also prevents the stress of startup loads, further reducing the mechanical strain on the compressor.

4. Quieter Operation:

   - **Low Sound Levels**: The compressor and the indoor unit's fan can operate at lower speeds, which significantly reduces the noise levels compared to traditional systems. This makes inverter mini-splits ideal for installation in noise-sensitive environments like bedrooms, offices, and libraries.

5. Better for the Environment:

   - **Reduced Greenhouse Gas Emissions**: By consuming less power, inverter mini-splits contribute to lower greenhouse gas emissions, aligning better with environmental sustainability goals.

- Compatibility with Low Global Warming Potential (GWP) Refrigerants: Many inverter mini-split systems are designed to be compatible with newer, more environmentally friendly refrigerants, which have lower global warming potentials.

6. Cost-Effective Operation:

- Lower Operating Costs: The increased efficiency of inverter systems often results in lower energy costs over the system's lifetime, offsetting the initial higher purchase price.

- Eligibility for Rebates and Incentives: Many regions offer rebates and incentives for installing energy-efficient systems like inverter mini-splits, providing additional financial benefits to consumers.

3. Comparison Between Inverter Mini-Splits And Traditional HVAC Systems

Understanding the differences between inverter mini-splits and traditional HVAC systems is crucial for anyone considering their options for heating and cooling solutions. Here's a detailed comparison highlighting the distinctions and relative advantages of each technology:

1. Operation Mechanism:

  - **Inverter Mini-Splits**: Utilize inverter technology, which allows the compressor to adjust its speed and output continuously based on the thermal load. This variable-speed operation is more energy-efficient and provides more precise temperature control.

  - **Traditional HVAC Systems**: Often rely on fixed-speed compressors, which turn on and off repeatedly to maintain the desired temperature.

This results in higher energy consumption and more noticeable temperature fluctuations.

2. Energy Efficiency:

- **Inverter Mini-Splits**: Generally more energy-efficient due to their ability to operate at variable speeds, minimizing energy waste associated with cycling on and off.

- **Traditional HVAC Systems**: Less energy-efficient as the constant starting and stopping of the compressor consumes a significant amount of energy, leading to higher operational costs.

3. Cost:

- **Inverter Mini-Splits**: Typically more expensive upfront due to the advanced technology and higher installation flexibility. However, the operational costs are lower over time, often resulting in overall savings.

- **Traditional HVAC Systems**: Less expensive initially but can be more costly to operate and maintain over their lifespan due to lower energy efficiency and higher wear and tear.

4. Installation Flexibility:

- Inverter Mini-Splits: Offer great installation flexibility, requiring only a small hole for the conduit that connects the indoor and outdoor units. They can be ideal for retrofitting older buildings without existing ductwork.

- **Traditional HVAC Systems**: Installation generally requires extensive ductwork, which can be disruptive and expensive, especially in buildings not originally designed for ducted systems.

5. Maintenance and Durability:

- **Inverter Mini-Splits**: Typically experience less wear and tear due to their smoother operational characteristics, potentially leading to a longer lifespan and lower maintenance costs.

- **Traditional HVAC Systems**: The frequent on-off cycling can lead to quicker component wear and higher maintenance needs.

6. Noise Levels:

- **Inverter Mini-Splits**: Operate more quietly, especially since the compressor can run at lower speeds most of the time.

- **Traditional HVAC Systems**: Tend to be noisier due to the abrupt starting and stopping of the compressor.

7. Climate Suitability:

- **Inverter Mini-Splits**: Highly effective in a wide range of climates due to their ability to adjust output precisely to match indoor conditions. They can maintain comfort even when outdoor temperatures are extremely high or low.

- **Traditional HVAC Systems**: May struggle to maintain efficiency in extreme climates, particularly when frequent cycling is required to keep up with demand.

8. Environmental Impact:

  - **Inverter Mini-Splits**: Generally considered more environmentally friendly due to higher energy efficiency and compatibility with newer, more eco-friendly refrigerants.

  - **Traditional HVAC Systems**: Often use older refrigerant types that are less environmentally friendly and less efficient, contributing to higher greenhouse gas emissions.

Overall, while traditional HVAC systems remain a viable option for many applications, inverter mini-splits offer numerous advantages in terms of efficiency, comfort, and environmental impact. These benefits make them an increasingly popular choice for both residential and commercial heating and cooling solutions.

Chapter 2: Basic Components And Functionality

1. Description Of Key Components (Outdoor Unit, Indoor Unit, Remote Control, Etc.)

Inverter mini-split systems consist of several essential components that work together to provide efficient heating and cooling. Understanding these components is crucial for anyone looking to install, maintain, or troubleshoot a mini-split system. Here are the main components of a typical inverter mini-split system:

1. Outdoor Unit (Compressor/Condenser):

- **Function**: The outdoor unit houses the compressor, which is responsible for pumping refrigerant through the system. It also contains the condenser coil where the refrigerant releases the heat it has absorbed from the indoor air.

- **Components**: Key components include the compressor, condenser coil, fan, and various control electronics. The compressor in an inverter mini-split is capable of varying its speed to match the cooling or heating demand more precisely.

- **Location**: Typically installed outside the building or on the roof, and must be placed in a location where airflow is not obstructed.

2. Indoor Unit (Evaporator):

- **Function**: The indoor unit contains the evaporator coil, which cools or heats the air as the refrigerant absorbs heat from it. This unit also includes a fan that circulates air over the evaporator coil and distributes the conditioned air throughout the room.

- **Components**: Includes the evaporator coil, a fan, air filter, and drainage components that handle condensation.

- **Types and Styles**: Comes in various forms such as wall-mounted, ceiling-mounted, floor-standing, or concealed duct units, allowing for flexible installation options depending on the room's aesthetics and functional requirements.

3. Refrigerant Lines:

- **Function**: Connect the indoor and outdoor units, allowing the refrigerant to circulate between them.

- **Components**: Consist of two lines - a smaller liquid line and a larger suction line. These lines are insulated to maintain efficiency and prevent energy loss.

- **Material**: Typically made of copper for its durability and high thermal conductivity.

4. Control System (Remote Control and Thermostat):

- **Function**: Allows users to set and adjust the temperature, change the mode of operation (heating, cooling, dehumidifying), and control other features such as fan speed and timer functions.

- **Components**: Most mini-splits come with a wireless remote control. Some models also offer wall-mounted thermostats or smart thermostat options that can be controlled via smartphone apps.

- **Features**: Advanced systems may include features like sleep modes, energy-saving modes, and air quality monitoring.

5. Power Supply:

- **Function**: Provides electrical power to the system. Mini-splits generally require a dedicated power line from the main electrical panel.

- **Components**: Includes wiring, circuit breakers, and sometimes a disconnect switch located near the outdoor unit for safety and maintenance purposes.

6. Drainage System:

- **Function**: Removes condensation produced during the cooling process from the indoor unit.

- **Components**: Consists of a drain pan and a drain line. Proper installation is crucial to prevent water leakage and potential damage to the property.

7. Accessories and Optional Components:

- **Function**: Enhance the functionality, aesthetics, or installation flexibility of the mini-split system.

- **Examples**: Includes line set covers, condensate pumps (for systems where gravity drainage is impractical), and specialized mounting brackets for specific installation scenarios.

2. How Inverter Technology Works

Inverter technology in mini-split systems represents a significant innovation in the control of HVAC systems, offering a smarter and more energy-efficient way to manage heating and cooling. This section explains how inverter technology works within the framework of a mini-split system and highlights the advantages it brings in terms of performance and energy consumption.

1. Basic Principle:

- **Variable Speed Compressor**: At the core of inverter technology is a variable-speed compressor that can adjust its speed based on the heating or cooling demand of the indoor environment. Unlike traditional compressors that turn on and off at full capacity, an inverter-driven compressor ramps up and down smoothly, maintaining a constant temperature with greater efficiency.

2. Modulation of Power:

- **Power Conversion**: Inverter systems start with converting the incoming AC (Alternating Current) electrical power to DC (Direct Current) using a rectifier. This DC power is then used to control the speed of the compressor with precision.

- **Frequency Adjustment**: An electronic controller adjusts the frequency of the electrical signal being sent to the compressor. By modifying the frequency, the controller can adjust the speed of the compressor motor. Higher frequencies make the compressor run faster, and lower frequencies slow it down.

3. Real-Time Adjustment:

- **Sensing and Feedback**: The inverter system continuously monitors conditions such as indoor temperature, outdoor temperature, and set temperature on the thermostat. Using this data, it calculates the required compressor speed to maintain the desired comfort level efficiently.

- **Continuous Operation**: Instead of shutting off entirely, the compressor operates continuously but at varying speeds. This continuous operation avoids the energy spikes associated with starting up the compressor, leading to more efficient energy use and reduced wear on system components.

4. Enhanced Comfort and Efficiency:

- **Precise Temperature Control**: The ability to adjust compressor speed allows the system to maintain a more consistent indoor temperature by making minor adjustments to the compressor's output. This precision avoids the typical temperature fluctuations associated with non-inverter systems and enhances overall comfort.

- **Energy Saving**: By operating at exactly the speed needed to maintain the set temperature, rather than cycling on and off, inverter mini-splits consume significantly less energy. This results in lower electricity bills and a reduced environmental impact.

5. Operational Advantages:

- **Rapid Response**: When cooling or heating needs to be increased rapidly, such as when initially turning on the unit or after opening a door, the inverter compressor can operate at higher speeds to quickly adjust the indoor climate to the desired temperature.

- **Quiet Operation**: The compressor and the fan can operate at lower speeds for the majority of the time, which significantly decreases the noise levels compared to traditional HVAC systems.

In summary, inverter technology enhances the functionality of mini-split systems by allowing for variable-speed control of the compressor. This leads to superior energy efficiency, more precise temperature control, quieter operation, and longer system life due to reduced mechanical stress. This technology represents a paradigm shift in how climate control is managed in residential and commercial settings, providing significant benefits over traditional fixed-speed HVAC systems.

3. Understanding The Refrigeration Cycle In Mini-Splits

The refrigeration cycle in a mini-split system is fundamental to its ability to heat and cool an environment efficiently. This cycle involves the continuous movement and phase change of refrigerant through various components of the system. Here's a breakdown of how the refrigeration cycle works in a mini-split system, particularly focusing on the flow and transformation of the refrigerant.

1. Compression:

   - **Starting Point**: The cycle begins when the refrigerant enters the compressor as a low-pressure gas. The compressor, driven by inverter technology for variable speed control, compresses this gas, raising its pressure and temperature.

- **Result**: The high-pressure, high-temperature gas then exits the compressor and moves towards the condenser.

2. Condensation:

  - **Location:** The hot refrigerant gas flows through the condenser coil, which is located in the outdoor unit.

  - **Process**: As the gas passes through the condenser, the outdoor fan blows ambient air over the coils, removing heat from the refrigerant. This causes the refrigerant to cool and change from a gas to a liquid state while still under high pressure.

  - **Outcome**: This phase change releases a significant amount of heat to the outside air, effectively cooling the refrigerant.

3. Expansion:

  - **Valve**: After leaving the condenser, the high-pressure liquid refrigerant flows through an expansion valve (or capillary tube) where its pressure abruptly drops. This sudden reduction in pressure lowers its temperature further.

   - **Effect**: The expansion valve controls the flow of refrigerant into the evaporator, ensuring that it enters at the correct rate to optimize the cooling effect.

4. Evaporation:

   - **Location**: The now cold, low-pressure liquid refrigerant enters the evaporator coil, located in the indoor unit.

   - **Process**: As indoor air is blown across the evaporator coil by the fan, the refrigerant absorbs heat from the air. This heat exchange causes the refrigerant to boil and turn back into a vapor, cooling the air as it passes through the coils.

   - **Result**: The cooled air is then circulated back into the room, while the refrigerant, having absorbed the heat, returns to the compressor as a low-pressure vapor to begin the cycle anew.

5. Heat Pump Functionality:

  - **Reversing Valve**: In heating mode, a device called a reversing valve is used to switch the direction of the refrigerant flow. This reversal transforms the function of the condenser and evaporator: the indoor coil becomes the condenser, releasing heat into the room, while the outdoor coil acts as the evaporator, extracting heat from the outside air.

  - **Versatility**: This dual functionality is what allows mini-split systems to provide both heating and cooling, making them an efficient choice for all-season climate control.

Understanding this refrigeration cycle is crucial for anyone involved in the installation, maintenance, or troubleshooting of mini-split systems, as it underpins the operation of the entire system. The cycle's efficiency is enhanced by inverter technology, which optimizes compressor speeds to adapt to the thermal demands of the space, thereby reducing energy consumption and improving overall system performance.

Chapter 3: Installation Guidelines

1. Pre-Installation Considerations (Sizing, Location, Etc.)

Proper installation is critical for ensuring that a mini-split system operates efficiently and effectively. Before installing an inverter mini-split system, several key considerations must be addressed to ensure optimal performance and compliance with local building codes. Here are the primary pre-installation considerations:

1. Sizing the System:

   - **Load Calculation**: Correctly sizing the mini-split system is crucial. A system that is too large will cycle on and off too frequently, reducing efficiency and increasing wear. Conversely, a system that is too small will struggle to adequately heat or cool the space, leading to increased energy consumption and inadequate climate control. Load calculations should consider room dimensions, window size and orientation, insulation levels, and local climate.

- **Good Assessment**: It is recommended to have a good HVAC technician perform a detailed load calculation using the Manual J calculation method, ensuring the selected system is appropriate for the space's specific needs.

2. Choosing the Right Location:

  - **Outdoor Unit**:

  - **Accessibility for Maintenance**: Ensure the outdoor unit is accessible for future maintenance and service. It should also be placed on a firm, level surface that can support its weight.

  - **Adequate Airflow**: The location should provide plenty of clearance around the unit to ensure adequate airflow and prevent recirculation of expelled air. It's also important to consider noise levels in relation to neighboring properties.

- **Protection from Elements**: While outdoor units are built to withstand weather conditions, positioning them in a location that minimizes exposure to direct sunlight, heavy winds, and rain can extend their lifespan and improve efficiency.

- **Indoor Unit:**

- **Wall Strength**: The chosen wall must be strong enough to hold the weight of the indoor unit.

- **Height from the Floor**: Typically, indoor units are installed 6-8 feet above the floor to ensure optimal air distribution and prevent cold or warm air from being wasted near the floor.

- **Away from Direct Sunlight and Obstacles**: Avoid installing the indoor unit where it will be directly exposed to sunlight or obstructed by furniture, curtains, or other household items.

- **Central Location**: For best results, place the indoor unit in a central location to allow for even distribution of air throughout the intended area.

3. Consideration of Refrigerant Lines:

- **Length and Route**: The distance between the indoor and outdoor units should be minimized to reduce the potential for refrigerant loss and efficiency degradation. However, if a long pipe run is necessary, additional refrigerant may need to be added.

- **Proper Insulation**: All refrigerant lines must be properly insulated to prevent energy loss and condensation issues.

4. Electrical Requirements:

- **Power Supply**: Ensure that the electrical supply matches the specifications required by the mini-split system. Most systems require a 240V outlet, and some might need a dedicated circuit.

- **Compliance with Electrical Codes**: Installation should comply with local electrical codes, and it might be necessary to have an electrical inspector review the installation.

5. Drainage System:

- **Condensate Disposal**: Proper planning for condensate drainage is essential. The indoor unit produces condensate that must be drained away from the unit and the building to prevent water damage and maintain indoor air quality.

Addressing these pre-installation considerations will help ensure that the mini-split system is installed correctly and operates efficiently. Proper installation not only maximizes comfort but also extends the life of the system while minimizing energy consumption and maintenance needs.

2. Step-By-Step Installation Process

Installing an inverter mini-split system involves several detailed steps that must be followed carefully to ensure optimal performance and longevity. Below is a step-by-step guide to help you through the installation process:

1. Preparing the Installation Site:

  - **Clear the Area**: Ensure both the indoor and outdoor installation sites are clear of debris, dust, and any obstacles that could impede the installation process or operation of the units.

  - **Confirm Measurements**: Double-check the measurements for where the units will be placed, ensuring they meet the required specifications for clearance and airflow.

2. Mounting the Indoor Unit:

  - **Wall Bracket Installation**: Install the mounting bracket on the interior wall at the correct height (usually about 6-8 feet above the floor), ensuring it is level and securely fastened to the wall studs.

- **Hang the Indoor Unit**: Carefully hang the indoor unit on the mounting bracket. Ensure it is securely in place and properly leveled.

3. Installing the Outdoor Unit:

- **Prepare the Base**: If the outdoor unit will be placed on the ground, prepare a concrete or plastic pad to sit it on. This base should be level and secure. For roof installations, ensure the support frame is sturdy and capable of handling the unit's weight plus potential snow or debris accumulation.

- **Position the Outdoor Unit**: Place the outdoor unit on the prepared base, maintaining proper distance from walls or obstructions to ensure adequate airflow.

4. Connecting Refrigerant Lines:

- **Route the Lines**: Extend the refrigerant lines from the indoor unit to the outdoor unit. Ensure the lines are as short as possible and properly insulated.

   - **Vacuum the Lines**: Before connecting, use a vacuum pump to remove any contaminants and moisture from the refrigerant lines. This is critical for the longevity and efficiency of the system.

   - **Connect and Seal**: Attach the refrigerant lines between the indoor and outdoor units. Ensure all connections are tight and secure to prevent leaks.

5. **Electrical Connections:**

   - **Wire the Indoor Unit**: Connect the electrical wires from the indoor unit to the outdoor unit, following the wiring diagram provided by the manufacturer.

   - **Establish Power Supply**: Ensure the electrical connections comply with local codes, and if necessary, install a new circuit breaker or disconnect switch to handle the electrical load.

6. Install the Drainage System:

- **Connect Drain Line**: Attach the drain hose to the indoor unit and route it to a suitable drainage location.

- **Ensure Proper Slope**: Make sure the drain line has a consistent downward slope without any loops or sags where water could collect.

7. Test the System:

- **Check for Leaks**: Before fully starting the system, check all refrigerant connections for leaks using a leak detector.

- **Start the Unit**: Turn on the power to the system and test it to ensure it operates correctly. Check the functioning of all modes (cooling, heating, and dehumidifying) and verify that the temperature controls and remote functions are working as expected.

8. Final Inspection and Cleanup:

- **Inspect Installation**: Make a final inspection of all installed components to ensure everything is secure and properly mounted.

- **Clean Up the Site**: Remove any installation debris and clean the area around both the indoor and outdoor units.

9. Customer Orientation:

- **Explain Operation**: It is good to explain the basic operation, maintenance, and troubleshooting of the system to the customer. Ensure they understand how to use the remote control and where the manuals are kept.

3. Common Mistakes To Avoid During Installation

Proper installation is crucial for the efficiency and longevity of inverter mini-split systems. Here are some common mistakes to avoid when installing these systems, ensuring optimal performance and avoiding future problems:

1. Incorrect Sizing of the System:

  - **Mistake**: Installing a mini-split system that is either too large or too small for the area it needs to serve. An oversized system will cycle on and off too frequently, which can lead to poor humidity control and increased wear. Conversely, an undersized system will struggle to effectively heat or cool, leading to overworking the unit.

  - **Prevention**: Always perform a detailed load calculation (Manual J) before choosing the system size to ensure it perfectly matches the space requirements.

2. Poor Location Choice for Indoor and Outdoor Units:

- **Mistake**: Placing the indoor unit in a location where furniture or other obstacles can obstruct airflow. For the outdoor unit, poor placement could be areas prone to high debris accumulation or lacking adequate clearance for air circulation.

- **Prevention**: Ensure the indoor unit is installed in a location with unobstructed air flow and the outdoor unit is placed in an area with sufficient space around it to facilitate proper air intake and heat dissipation.

3. Inadequate Clearance around Units:

- **Mistake**: Not providing enough clearance around the indoor or outdoor units, which can restrict airflow and reduce system efficiency.

- **Prevention**: Follow manufacturer guidelines for clearance around all sides of both the indoor and outdoor units to ensure effective operation.

4. Incorrect Refrigerant Line Length and Installation:

- **Mistake**: Using refrigerant lines that are too long, which can decrease efficiency, or too short, which may place strain on connections. Improperly insulated lines can also lead to energy losses.

- **Prevention**: Cut refrigerant lines to the exact needed length and ensure proper insulation. Avoid unnecessary bends and maintain the minimum bend radius recommended by the manufacturer.

5. Inadequate Vacuuming of Refrigerant Lines:

- **Mistake**: Not vacuuming the refrigerant lines properly before activation, leading to moisture and air remaining in the lines, which can cause system inefficiencies and potential damage.

- **Prevention**: Always use a proper vacuum pump to evacuate the lines thoroughly before filling them with refrigerant. This process removes any moisture and air trapped inside, which is crucial for the longevity and efficiency of the system.

6. Poor Electrical Connections:

  - **Mistake**: Loose or incorrect electrical connections can lead to electrical hazards or system malfunction.

  - **Prevention**: Ensure all electrical wiring is properly connected according to the wiring diagram provided by the manufacturer. Use appropriate cable sizes and secure connections tightly. It is often advisable to have a certified electrician review the electrical connections if you are not experienced.

7. Overlooking Condensate Drainage:

  - **Mistake**: Improper installation of the condensate drainage system can lead to water leaks and potential damage to property.

  - **Prevention**: Ensure that the condensate drain line is properly sloped away from the indoor unit and that it discharges in an appropriate location. Check for any kinks or blocks that might impede water flow.

8. Skipping the Leak Test:

- **Mistake**: Not testing the system for refrigerant leaks after installation.

- **Prevention**: After installing the refrigerant lines, perform a thorough leak test using a suitable leak detector. Any leaks should be fixed before starting the system.

By avoiding these common mistakes, installers can ensure a smooth and efficient setup that maximizes the performance and lifespan of the inverter mini-split system. Proper installation not only ensures optimal operation but can significantly reduce the need for future maintenance and repairs.

Chapter 4: Operating Inverter Mini-Splits

1. Detailed Guide On Operating The System

Learning to operate your inverter mini-split system efficiently can greatly enhance your comfort and save on energy costs. This chapter provides a detailed guide on how to operate these systems effectively.

1. Understanding the Control Panel and Remote Control:

  - **Key Features**: Familiarize yourself with the main functions on your mini-split's control panel or remote control. Common features include power on/off, temperature settings, fan speed control, and mode selection (cool, heat, dry, fan-only).

  - **Temperature Settings**: For optimal energy efficiency, set the thermostat to a comfortable yet economical temperature. Typically, 78°F (26°C) for cooling and 68°F (20°C) for heating are recommended.

- **Timer Functions**: Learn to use the timer for setting the system to turn on or off automatically. This can be useful for managing energy usage without sacrificing comfort, such as cooling or heating the space before you arrive home.

2. Selecting Operating Modes:

- **Cool Mode**: Use this mode for cooling your space. Set the desired temperature and adjust the fan speed as needed.

- **Heat Mode**: In colder months, switch to heat mode. The inverter technology will work to maintain the set temperature efficiently.

- **Dry Mode**: This mode is useful in humid conditions. It helps remove moisture from the air without significant cooling, making the space feel more comfortable.

- **Fan-Only Mode**: Use this mode to circulate air without heating, cooling, or dehumidifying. It's great for mild weather days or when fresh air is needed in the room.

3. **Adjusting Fan Speeds:**

   - **Options**: Most units offer multiple fan speeds. Higher speeds can be used for quick temperature changes, while lower speeds are better for maintaining a steady temperature with quieter operation.

   - **Automatic Fan Speed**: Many systems include an auto fan setting, which adjusts the fan speed based on the difference between the ambient room temperature and the set temperature on the thermostat.

4. **Using Advanced Features:**

   - **Energy-Saving Mode**: This feature, often marked as "Eco" mode, optimizes the system for energy efficiency by adjusting the temperature setting slightly to reduce power consumption.

- **Sleep Mode**: Activating sleep mode will gradually adjust the room temperature during the night for comfortable sleeping conditions and energy efficiency.

- **Wi-Fi/Smart Features**: If your unit is equipped with Wi-Fi capabilities, take advantage of controlling it through a mobile app. This can include adjusting settings remotely, monitoring energy usage, and receiving maintenance reminders.

5. Regular Monitoring and Adjustments:

- **Check Regularly**: Regularly check the system's settings to ensure it is operating as expected. Small adjustments can often prevent larger issues and maintain system efficiency.

- **Adapt to Weather Changes**: Be proactive in adjusting your system's settings based on current weather conditions and forecasted temperatures. This responsiveness can help maintain comfort and optimize energy usage.

6. Troubleshooting Basic Issues:

- **Common Issues**: Be familiar with common issues such as the unit not turning on, not heating or cooling effectively, producing strange noises, or emitting unusual odors. Check the user manual for troubleshooting tips and know when to call for professional help.

- **Resetting the System**: Sometimes, simply resetting the mini-split (turning it off at the circuit breaker and waiting before turning it back on) can resolve minor operational issues.

2. Setting Up And Understanding The Control Panel And Remote Operations

Inverter mini-splits come equipped with control panels and remote controls that allow users to adjust settings conveniently to manage indoor climate effectively. Here's how to set up and understand these controls for optimal operation:

1. Understanding the Remote Control:

- **Power Button**: Typically, the power button turns the system on and off. It may also be used to wake up the remote if it has a screen that goes to sleep.

- **Temperature Controls**: Use these buttons to set the desired temperature. The remote may display the set temperature and, depending on the model, the current room temperature.

- **Mode Selection**: The mode button allows you to cycle through different operating modes such as cool, heat, dry (dehumidify), and fan-only. Each mode may have an icon or a specific light indicator.

- **Fan Speed**: Adjusts the fan speed among low, medium, high, or auto. In auto mode, the system selects the fan speed based on the current room conditions and the set temperature.

- **Timer/Setback**: This function can be used to program the system to turn on or off at specific times, which is useful for energy savings and convenience.

2. Advanced Features on the Remote Control:

- **Eco Mode**: Engages energy-saving settings that slightly adjust the temperature setting to reduce power consumption.

- **Sleep Mode**: Gradually changes the temperature during the night to ensure comfort and energy efficiency while sleeping.

- **Swing or Louver Control**: Directs the movement of the air vents (louvers) to control airflow direction. This can often be set to swing (move automatically) or fixed in a certain direction.

- **Quiet Mode**: Reduces the operation noise of the indoor unit by lowering the fan speed, useful for night time or during quiet activities.

3. Control Panel Setup:

- **On-Unit Controls**: Some mini-splits also include a basic control panel on the indoor unit itself. This might be used for basic operations like turning the unit on/off or changing the temperature if the remote is not available.

- **Reset Button**: Most units have a reset button on the control panel or inside the unit, which can be used to reset the system in case of operational issues or after a filter change.

- **Indicator Lights**: Various lights may indicate the mode of operation, whether the unit is in defrost mode, and other statuses. It's important to refer to the user manual to understand what each light means.

4. Setting Up the System:

  - **Initial Setup**: When setting up the unit for the first time, ensure all factory settings are appropriate for your needs. This includes checking the default temperature settings, fan speeds, and program settings.

  - **Programmable Features**: Utilize the timer and programmable features to set daily or weekly heating and cooling schedules based on your routine, which can significantly enhance comfort and efficiency.

5. Regular Maintenance Reminders:

  - **Filter Check/Clean Indicator**: Many systems include a reminder light or notification on the remote or control panel indicating when it's time to check or replace the air filter.

- **Error Codes**: Should the system experience any faults, error codes may be displayed on the remote control's screen or the unit's LED panel. Refer to the user manual to understand these codes and determine if a professional service call is needed.

Understanding how to effectively use the control panel and remote control can greatly enhance your experience with your inverter mini-split system. Familiarity with these elements allows for tailored comfort settings, efficient operation, and troubleshooting minor issues, contributing to the overall satisfaction with the system's performance.

3. Tips For Optimal Efficiency And Comfort

Operating your inverter mini-split system efficiently not only extends the life of the equipment but also ensures maximum comfort and minimizes energy costs. Here are several tips to help you achieve optimal efficiency and comfort with your mini-split system:

1. Optimal Temperature Settings:

   - **Cooling**: Set the thermostat to 78°F (26°C) when you are at home and active. This temperature is generally comfortable for most people and is energy-efficient.

   - **Heating**: During winter, set the thermostat to 68°F (20°C) when you're awake and lower it while you're asleep or away from home.

   - **Avoid Extreme Temperature Settings**: Setting the thermostat to extreme temperatures doesn't cool or heat the room faster and can lead to unnecessary energy consumption.

2. Use Programmable Timers:

  - **Automate Operation**: Use the timer function to turn the system on or off based on your daily schedule, which avoids running the system when it's not needed and can significantly reduce energy usage.

3. Take Advantage of Inverter Technology:

  - **Stable Operation**: Allow the system to run at a steady state as much as possible. Inverter systems are designed to adjust their output to the needs of the space efficiently, which means they use less energy maintaining a stable temperature compared to reaching a temperature.

  - **Gradual Temperature Changes**: When needing to change the set temperature, do it gradually. Sudden large changes force the system to work harder, reducing efficiency.

4. Maintain Adequate Airflow:

- **Interior Airflow**: Keep the area around the indoor unit clear of furniture and other obstructions to maintain good air circulation, ensuring the unit can effectively distribute air throughout the room.

- **Outdoor Unit Clearances**: Make sure the outdoor unit has sufficient clearance from plants and structures to allow for proper air intake and heat dissipation.

5. Regular Maintenance:

- **Clean Filters Regularly**: Check and clean the air filters monthly during peak usage periods. A clean filter ensures efficient operation and good indoor air quality.

- **Seasonal Checks**: Before the start of the heating or cooling season, inspect and clean the components of your mini-split, such as the coils and condensate lines.

6. Utilize Eco and Sleep Modes:

- **Eco Mode**: Engage Eco mode during times when slight fluctuations in temperature are acceptable. This mode optimizes the system's operation to improve energy efficiency without a significant impact on comfort.

- **Sleep Mode**: Use sleep mode overnight, which gradually adjusts the room temperature to prevent the system from working too hard and to enhance sleeping comfort.

7. Optimal Fan Settings:

- **Auto Fan Speed**: Use the auto fan setting for most situations. The system will automatically adjust the fan speed based on the current and target temperatures, which is more energy-efficient than keeping the fan on a constant high or low setting.

- **Directional Airflow**: Adjust the direction of the airflow (if your unit allows it) so that air is not blowing directly on occupants. This can increase comfort and allow you to set the thermostat at a more energy-efficient temperature.

8. Block Sunlight and Reduce Drafts:

   - **During Summer**: Use curtains or blinds to block sunlight during the hottest part of the day to help keep your space cooler.

   - **During Winter**: Ensure windows and doors are properly sealed to prevent drafts that can make the space feel colder and cause the system to work harder.

Chapter 5: Maintenance Essentials

1. Routine Maintenance Schedule

Regular maintenance is crucial to ensuring that your inverter mini-split system operates efficiently and effectively throughout its lifespan. Establishing a routine maintenance schedule can prevent common issues, enhance performance, and extend the system's durability. Here is a recommended routine maintenance schedule:

1. Monthly Checks and Maintenance:

  - Air Filters:

  - Cleaning: Clean the air filters every month, especially during high-use seasons. Remove the filters according to the manufacturer's instructions, use a vacuum to remove dust, and wash them with mild soap and water if necessary. Ensure they are completely dry before reinstalling.

  - Replacement: Replace filters if they are damaged or excessively worn, typically every 3-6 months depending on usage and environmental conditions.

2. Seasonal (Every 3-6 Months) Maintenance:

- **Indoor and Outdoor Coils:**

  - **Cleaning**: Check and clean the evaporator and condenser coils every season. Dirt and debris can insulate the coils, reducing efficiency and cooling capacity.

- **Condensate Drain:**

  - **Check and Clear**: Ensure that the condensate drain line is clear of obstructions. Flush it periodically to prevent clogs that can lead to water leaks and potential damage to your property.

3. Annual Maintenance:

- **Professional Inspection:**

  - **Overall System Check**: Have a professional HVAC technician inspect the system annually. This inspection should include checking refrigerant levels, testing for leaks, inspecting electrical connections, and ensuring that all components are functioning properly.

- **Ductless System Specific Checks:**

  - **Check Fan and Motor**: Ensure that the fan and motor are operating efficiently and quietly. Lubricate bearings and replace the fan if it shows signs of wear or damage.

  - **Inspect Remote Control**: Check the remote control for functionality and replace batteries as needed.

4. Biannual (Every Two Years) Maintenance:

- **Deep Cleaning**:

  - **Professional Coil Cleaning**: Although regular user maintenance is essential, professional cleaning of the coils every two years can help maintain optimal performance, especially in environments prone to dust or pollution.

- **System Calibration:**

- **Check Calibration**: Ensure that the thermostat and control systems are accurately calibrated to reflect actual temperatures and operating conditions. Misalignment can lead to inefficiency and discomfort.

5. As Needed Maintenance:

- **Noise and Vibration Checks:**

- **Inspect for Unusual Noises**: Listen for any abnormal noises during operation, such as rattling or buzzing, which can indicate loose components or obstructions. Address these issues promptly to avoid further complications.

- **Update Settings**:

- **System Updates**: If your system features software controls or smart functionality, ensure that the firmware or software is kept up to date according to the manufacturer's recommendations.

Maintenance Record Keeping:

 - **Log All Maintenance** Activities: Keep a record of all maintenance activities, including what was done, who performed the maintenance, and when it was completed. This documentation can be helpful for warranty purposes and for scheduling future maintenance.

Adhering to this routine maintenance schedule will help ensure that your inverter mini-split system remains in optimal condition, providing reliable and efficient heating and cooling for years to come. Regular maintenance not only prevents unexpected breakdowns but also ensures that the system operates at peak efficiency, reducing energy costs and environmental impact.

2. Cleaning And Replacing Filters

One of the most critical maintenance tasks for ensuring the efficient operation of your inverter mini-split system is the regular cleaning and occasional replacement of its air filters. Here's a detailed guide on how to properly clean and replace the filters:

1. Cleaning Air Filters:

  - **Frequency**: Clean the filters every month, especially during periods of high usage. If you live in a dusty area or have pets, you might need to clean the filters more frequently.

  - **Steps for Cleaning:**

    1. **Turn Off the Unit**: Always ensure the unit is turned off before attempting any maintenance tasks.

    2. **Remove the Filters**: Open the front panel of the indoor unit. The filters are typically located

just behind this panel and can be easily slid out or unclipped.

3. Vacuum the Filters: Use a vacuum cleaner with a soft brush attachment to gently remove the bulk of the dust and debris from both sides of the filters.

4. Wash the Filters: For a deeper clean, rinse the filters under running water. For stubborn dirt, use a mild detergent and gently scrub with a soft brush. Avoid using hot water or strong chemicals, which can damage the filter material.

5. Dry Thoroughly: After washing, shake off excess water and let the filters dry completely in a shaded area. Avoid direct sunlight or heat sources, which can warp or degrade the filter material.

6. Reinstall the Filters: Once completely dry, reinstall the filters in the unit. Ensure they are properly seated and that the front panel is securely closed.

2. Replacing Air Filters:

- **When to Replace**: Replace the air filters if they are damaged, excessively worn, or if they still seem clogged after cleaning. It's generally recommended to replace mini-split air filters every 6 to 12 months, depending on usage and environmental conditions.

- **Steps for Replacement:**

  1. Purchase the Correct Filters: Obtain replacement filters that are specifically designed for your mini-split model. Using the correct filters ensures optimal efficiency and prevents damage to the unit.

  2. Remove Old Filters: Follow the same steps as for cleaning to remove the old filters.

  3. Install New Filters: Slide the new filters into place, ensuring they fit securely and that no gaps are left around the edges.

4. Dispose of Old Filters Properly: Dispose of the old filters according to local regulations, as they may contain dust and other allergens.

3. Best Practices:

- **Regular Checks**: Even if you do not think the filters are dirty, it's good practice to check them monthly. This helps prevent unexpected issues and maintains air quality.

- **Maintain Airflow**: Clean and correct positioning of filters ensure optimal airflow, which is crucial for the efficiency of your system. Reduced airflow can cause the system to work harder, increasing energy consumption and reducing its lifespan.

By following these guidelines for cleaning and replacing the filters of your inverter mini-split system, you can maintain its efficiency, prolong its operational life, and ensure a healthier indoor environment. Regular filter maintenance is a simple yet effective way to enhance the overall performance of your heating and cooling system.

3. Checking And Maintaining Refrigerant Levels

Refrigerant is the lifeblood of any air conditioning system, including inverter mini-split systems. It's essential for the heat transfer process that cools and heats your space. Proper refrigerant levels are crucial to ensure efficient operation and prevent damage to the system. Here's how to check and maintain refrigerant levels:

1. Importance of Correct Refrigerant Levels:

  - **Efficiency**: Proper refrigerant levels help maintain the efficiency of your mini-split system. Low refrigerant levels can lead to higher energy consumption and reduced cooling or heating performance.

  - **System Health**: Low or leaking refrigerant can cause the compressor to overheat and fail prematurely. Conversely, too much refrigerant can lead to high pressure in the system, which may damage components.

2. Signs of Refrigerant Issues:

- **Inadequate Cooling or Heating**: If the mini-split isn't cooling or heating effectively, it might be due to insufficient refrigerant.

- **Ice Build-up**: Ice forming on the refrigerant lines or the outdoor unit can indicate low refrigerant levels.

- **Hissing or Bubbling Noises**: Sounds like hissing or bubbling from the refrigerant lines could suggest a leak.

- **Increased Energy Bills**: An unexplained increase in energy bills might be due to the system working harder to compensate for the lack of refrigerant.

3. Checking Refrigerant Levels:

- **Inspection**: Refrigerant levels should be checked. The process involves using gauges to measure the pressure in the system while it's running. The pressure readings are compared to the specifications recommended by the system's manufacturer.

4. Maintaining Refrigerant Levels:

- **Detecting and Repairing Leaks:**

- **Leak Detection**: Technicians may use electronic leak detectors, UV dyes, or bubble solutions to find where refrigerant is escaping.

- **Repairing Leaks**: Once a leak is detected, it must be repaired by soldering or replacing parts of the refrigerant circuit. This should be done before refilling the system with refrigerant.

- **Recharging Refrigerant**:

- **Correct Type and Amount**: Only a certified technician should recharge the system, using the type and amount of refrigerant specified by the manufacturer. This ensures compliance with regulations and maintains system performance.

- **Testing Post-Recharge**: After recharging, the system should be tested to ensure it operates correctly and efficiently.

5. Environmental and Legal Considerations:

- **Regulations**: Handling of refrigerants is regulated due to their environmental impact, particularly their role in depleting the ozone layer and contributing to global warming.

- **Certification Requirements**: Only professionals with EPA (Environmental Protection Agency) certification should handle refrigerants due to the complexities and environmental regulations involved.

6. Preventative Maintenance:

- **Regular Inspections**: Schedule annual or bi-annual inspections with a professional to check refrigerant levels and overall system health.

- **Prompt Repairs**: Address any issues promptly to avoid the escalation of damage and to maintain system efficiency.

By adhering to these guidelines for checking and maintaining refrigerant levels, you can help ensure that your inverter mini-split system remains efficient, effective, and safe over its operational life. Regular professional maintenance is key to sustaining performance and avoiding costly breakdowns.

Chapter 6: Troubleshooting Common Issues

1. Diagnosing Common Problems

Inverter mini-split systems are highly efficient and reliable, but like any mechanical system, they can experience issues. Understanding how to diagnose common problems can help you quickly address them, often without needing professional assistance. Here are some typical problems you might encounter and how to troubleshoot them:

1. Not Cooling/Heating Effectively:

- **Check Air Filters**: Ensure that the air filters are clean. Dirty filters can restrict airflow, significantly reducing the system's efficiency.

- **Inspect Airflow**: Confirm that nothing is blocking the indoor or outdoor unit's airflow.

- **Check Thermostat Settings**: Make sure the thermostat is set to the correct mode (cooling or heating) and the desired temperature.

- **Look for Ice Build-up**: Ice on the coils or refrigerant lines can indicate issues such as low refrigerant levels or airflow problems.

- **Examine Refrigerant Levels**: Low refrigerant can cause poor cooling or heating. This requires a professional to check and refill if necessary.

2. Unusual Noises:

- **Rattling**: Often caused by loose screws or panels. Tighten any loose elements you find.

- **Hissing or Bubbling**: This could indicate a refrigerant leak. Due to the complexity and environmental concerns, it's best to call a professional to address refrigerant issues.

- **Grinding or Squealing**: These sounds may come from the fan or motor bearings. If lubrication doesn't solve the problem, parts may need to be replaced by a technician.

3. Water Leaks:

- **Check Condensate Drain**: Ensure the drain is not clogged. Flush the line to clear any blockages.

- **Inspect Installation**: Improper installation can cause water to drain improperly. Make sure all units are level and that drainage flows away from the unit.

- **Examine Pump (if applicable):** If your system uses a condensate pump, ensure it is functioning correctly and is not clogged.

4. System Does Not Turn On:

- **Power Supply**: Check if the circuit breaker has tripped or if there is a blown fuse.

- **Remote Control**: Verify that the remote control has working batteries.

- **Inspect Wiring**: Loose wiring can also prevent the system from powering up. Check connections, but avoid handling wiring if you are not experienced—consult a professional.

5. Odors:

- **Cleanliness**: Musty smells can indicate mold or mildew within the system, usually due to a dirty filter or moisture in the unit. Cleaning the filter and ensuring proper drainage can help.

- **Burnt Smells**: Electrical odors may suggest a malfunctioning component. Turn off the system and consult a professional.

6. System Frequently Turns Off and On:

- **Oversized Unit**: If the mini-split is too large for the space, it may cycle too frequently, which is inefficient and wears the system down faster.

- **Thermostat Issues**: Ensure the thermostat is not in direct sunlight or near other heat sources, as this can cause incorrect temperature readings.

7. Remote Control Malfunctions:

- **Battery Replacement**: Simple but often overlooked, replacing the remote's batteries can resolve many issues.

- **Signal Reception**: Make sure there is a clear path between the remote and the unit's sensor.

2. Step-By-Step Troubleshooting Procedures

Troubleshooting your inverter mini-split system can help identify and resolve issues that may affect its performance. Here are detailed steps to diagnose and address common problems:

1. System Not Cooling or Heating Properly:

   - **Step 1**: Check the thermostat settings to ensure the system is on the correct mode (heat or cool) and set to a reasonable temperature.

   - **Step 2**: Inspect and clean the air filters if dirty, as clogged filters can significantly impair performance.

   - **Step 3**: Verify that there are no obstructions blocking the indoor and outdoor unit's airflow.

   - **Step 4:** Look for signs of ice on the coils or refrigerant lines, which can indicate low refrigerant levels or airflow issues.

- **Step 5**: If suspected, check refrigerant levels and system pressure.

2. Unusual Noises (Rattling, Grinding, Hissing):

- **Step 1**: Identify the type of noise, which can help pinpoint the issue (rattling usually indicates loose parts; grinding may suggest motor or bearing issues; hissing might be a refrigerant leak).

- **Step 2**: Secure any loose panels or screws if the noise is a rattling sound.

- **Step 3**: If grinding, check the condition of the fan and motor bearings. Lubricate if necessary, or plan for replacement.

- **Step 4**: For hissing noises, avoid handling it yourself as it could be a refrigerant leak. Call a professional.

3. Water Leaking from the Unit:

- **Step 1:** Check the condensate drain for blockages. Clear any debris or buildup in the drain line.

- **Step 2**: Ensure the unit is properly leveled. An improperly leveled unit can cause improper drainage.

- **Step 3**: Inspect the condensate pump (if installed) for proper operation. Clean or replace as needed.

4. System Fails to Power On:

- **Step 1**: Verify that the unit is receiving power by checking the circuit breakers and ensuring no switches are turned off.

- **Step 2**: Check the remote control batteries and replace if dead.

- **Step 3**: Examine visible wiring connections for looseness or damage. Consult a technician for repairs as dealing with electrical components can be hazardous.

5. System Cycles On and Off Too Frequently:

- **Step 1**: Check if the system may be too large for the space, which can cause short cycling.

- **Step 2**: Ensure the thermostat is not influenced by external heat sources like lamps, appliances, or direct sunlight.

- **Step 3**: Inspect the air filters and replace or clean them if they are dirty.

6. Detecting and Addressing Odors:

- **Step 1**: Identify the type of odor. Musty smells usually indicate mold or mildew; burnt odors suggest electrical issues.

- **Step 2**: Clean or replace the air filters and inspect the evaporator coil for cleanliness.

- **Step 3**: For persistent mold issues, clean the system more thoroughly or consider using a UV light treatment. For electrical smells, immediately turn off the system and consult a professional.

7. Remote Control Not Working:

- **Step 1**: Replace the batteries in the remote control.

- **Step 2**: Ensure the remote's signal is not being blocked and the sensor on the unit is clean and unobstructed.

- **Step 3**: Try resetting the remote, if possible, or consult the manual for specific troubleshooting related to the remote.

Chapter 7: Advanced Service Procedures

1. Technical Service Procedures

Inverter mini-split systems, like all sophisticated mechanical systems, may require advanced technical services beyond routine maintenance and basic troubleshooting. These advanced procedures typically involve critical components such as the compressor or the inverter board and should always be performed by qualified HVAC technicians. Here's an overview of some key advanced service procedures:

1. Compressor Replacement:

   - **Diagnosis**: Compressor failure can manifest through symptoms such as not cooling/heating, making unusual noises, or causing circuit breakers to trip frequently. Diagnosis should involve checking the compressor's motor resistance, inspecting for signs of refrigerant leakage, and testing the compressor's electrical windings.

- **Procedure:**

1. **Power Down**: Ensure the system is completely powered off and disconnected from the electrical supply.

2. **Recover Refrigerant**: Use a refrigerant recovery machine to safely remove refrigerant from the system.

3. **Remove Old Compressor**: Disconnect all electrical connections, unbolt the compressor from its mounting, and carefully remove it from the system.

4. **Install New Compressor**: Position the new compressor, secure it onto the mounting, reconnect electrical wiring, and ensure all seals are properly fitted.

5. **Vacuum and Recharge**: Perform a vacuum on the system to remove any air and moisture before recharging with refrigerant.

6. Test the System: Power up the system and monitor it for proper operation, checking for leaks and verifying that the compressor operates as expected.

2. Inverter Board Troubleshooting and Replacement:

- **Symptoms**: Problems such as the system not responding to controls, erratic behavior, or failure to start often trace back to issues with the inverter board.

- **Procedure**:

1. Safety Precautions: Power down the unit completely to avoid electrical shock.

2. Diagnostic Checks: Use multimeters and oscilloscopes to check for faulty components on the board, looking specifically at capacitors, resistors, and power transistors.

3. Remove Faulty Board: Once a fault is confirmed, disconnect all connections and carefully remove the faulty inverter board.

4. Install Replacement Board: Secure the new inverter board in place, reconnect all wiring, and ensure every connector is properly seated.

5. System Testing: Re-energize the system and conduct functional tests to ensure the new board is operational and the system responds correctly to all control inputs.

3. Advanced Refrigerant Leak Detection and Repair:

- **Detection Techniques**: Beyond basic bubble tests, use electronic leak detectors, ultraviolet dye, and infrared imaging to locate leaks, especially in concealed areas.

- **Repair Procedure**:

1. Identify Leak Points: Confirm the exact location of leaks.

2. Recover Refrigerant: Safely remove the refrigerant from the system.

3. Repair Leaks: Depending on the nature and location, this might involve brazing or replacing sections of the refrigerant circuit.

4. Pressure Test: Perform a pressure test using nitrogen to verify the integrity of the repair.

5. Vacuum and Recharge: Evacuate the system to remove air and moisture, then recharge with the correct type and amount of refrigerant.

6. Test for Efficiency: Check system operation to ensure it returns to optimal efficiency levels.

2. Tools And Equipment Needed For Advanced Servicing

Servicing advanced components of an inverter mini-split system, such as compressors, inverter boards, and refrigerant circuits, requires specialized tools and equipment. These tools ensure that servicing is done efficiently, safely, and in compliance with industry standards. Here's a list of essential tools and equipment needed for advanced servicing of mini-split systems:

1. Refrigerant Recovery Machine:

   - **Purpose**: Safely extract and store refrigerant from the system before repairs or component replacement.

2. HVAC Gauges:

   - **Purpose:** Measure the pressure of refrigerant within the system to diagnose issues and verify system performance after servicing.

3. Vacuum Pump:

- **Purpose**: Remove air and moisture from the refrigeration system, which is critical after opening the system for any repairs or before recharging refrigerant.

4. Refrigerant Scale:

- **Purpose**: Measure the exact amount of refrigerant added to or recovered from the system, ensuring accurate charging.

5. Leak Detector:

- **Purpose**: Detect refrigerant leaks. Advanced detectors can pinpoint very small leaks, which are critical to system efficiency and environmental safety.

6. Multimeter and Clamp Meter:

- **Purpose**: Measure electrical values such as voltage, current, and resistance. Essential for diagnosing electrical components, including inverter boards and compressors.

7. Oscilloscope:

- **Purpose**: Used for more in-depth electrical diagnostics, particularly useful in troubleshooting inverter boards.

8. Brazing and Soldering Kit:

- **Purpose**: Repair and secure refrigerant lines and other metal components. Includes torch, solder, flux, and safety equipment.

9. Torque Wrench:

- **Purpose**: Ensure that all bolts and fittings are tightened to the specified torque settings, critical for preventing leaks at connection points.

10. UV Dye Kit and UV Light:

- **Purpose**: Inject UV dye into the refrigerant circuit to detect leaks. The UV light makes the dye visible at leak points.

11. Digital Manifold Set:

- **Purpose**: An advanced tool for diagnosing system performance that provides digital readouts of system pressures, temperatures, and vacuum levels.

12. Nitrogen Regulator and Cylinders:

- **Purpose**: Used for pressure testing and purging air from the system during leak testing or repairs.

13. Thermographic Camera:

- **Purpose**: Detect thermal anomalies in the system, which can indicate issues like electrical faults, blocked lines, or uneven cooling.

14. Screwdrivers, Pliers, and Wrench Sets:

- Purpose: General hand tools for opening the unit, adjusting components, and replacing parts.

15. Safety Gear:

- **Purpose**: Personal protective equipment, including gloves, goggles, and potentially a respirator, to protect against potential hazards like refrigerant exposure or burns from hot components.

3. Safety Precautions And Best Practices

Working with inverter mini-split systems, especially during advanced servicing tasks, requires adherence to stringent safety precautions and best practices to ensure both technician safety and system integrity. Here's a detailed guide on the safety measures and best practices to follow:

1. Safety Precautions:

  - **Electrical Safety**: Always ensure the power to the system is completely turned off at the circuit breaker before starting any work. Use a multimeter to verify that no live power is present.

  - **Refrigerant Handling**: Refrigerants should be handled with care as they can be hazardous. Wear gloves and goggles to protect from potential splashes. Refrigerant exposure can cause frostbite or burns.

  - **Proper Ventilation**: When working with refrigerants or soldering components, ensure the area is well-ventilated to avoid inhalation of fumes, which can be harmful.

- **Fire Safety**: When brazing or soldering, keep a fire extinguisher nearby. Be aware of flammable materials and clear the area of any potential fire hazards.

- **Lifting Techniques**: Use proper lifting techniques and mechanical aids when moving heavy equipment like compressors to avoid back injuries.

- **Tool Safety**: Use tools appropriately and maintain them properly. Inspect tools for damage before use and replace any that are worn or defective.

2. Best Practices in System Handling:

- **Documentation**: Always have the manufacturer's service manual and wiring diagrams on hand. Understanding specific system requirements and configurations is crucial for effective troubleshooting and repairs.

- **Clean Work Area**: Keep the workspace clean and organized. This not only prevents accidents but also helps in keeping small parts from getting lost.

- **Systematic Approach**: Follow a systematic approach in diagnosing problems. This includes checking the simplest possibilities first before moving on to more complex diagnostics.

- **Use of Proper Tools**: Utilize the correct tools for each task. This not only makes the work safer but also protects the system from potential damage caused by the use of inappropriate tools.

- **Quality Replacement Parts**: Always use OEM (original equipment manufacturer) parts or those of equivalent quality. Using substandard parts can lead to more frequent failures and can void warranties.

3. Refrigerant-Specific Best Practices:

- **Legal Compliance**: Ensure compliance with all local, state, and federal regulations regarding refrigerant handling, including proper recovery, recycling, or disposal of refrigerants.

- **Leak Detection**: Regularly check for refrigerant leaks using appropriate methods. Promptly addressing leaks not only prevents environmental damage but also maintains system efficiency.

- **Accurate Recharging**: When refilling the refrigerant, ensure that the exact type and quantity specified by the manufacturer are used. Incorrect refrigerant type or charge levels can severely impact system efficiency and longevity.

4. Record Keeping:

- **Maintenance Logs**: Keep detailed records of all maintenance and repairs performed. This helps in tracking the history of the system, planning future maintenance, and troubleshooting recurring issues.

- **Warranty Documentation**: Maintain all records of replaced parts and service visits as these are often required for warranty claims.

Chapter 8: Case Studies

1. Real-World Examples Of Troubleshooting And Repair

Case studies are invaluable for understanding how theoretical knowledge is applied in practical situations, especially in complex fields like HVAC. This section presents real-world examples of troubleshooting and repair on inverter mini-split systems, illustrating the process and solutions used to resolve common issues.

Case Study 1: Inconsistent Cooling and Ice Buildup

Situation:

A residential inverter mini-split system began showing signs of inconsistent cooling in certain zones and ice formation on the refrigerant lines outside the house.

Troubleshooting Steps:

1. Visual Inspection: Initial inspection showed frost on the exterior refrigerant lines, suggesting a possible refrigerant issue.

2. Check Airflow: The technician checked the indoor unit's filters and found them to be clogged, restricting airflow.

3. Refrigerant Pressure Test: After cleaning the filters, the system was still not performing optimally, prompting a refrigerant pressure test. The test confirmed that the refrigerant levels were indeed low.

Solution:

The technician repaired a small leak detected at a refrigerant line connection, evacuated the system to remove any air and moisture, and then recharged it with the appropriate amount of refrigerant. Post-repair, the system's cooling efficiency was restored, and no further ice buildup was observed.

Case Study 2: System Not Powering On

Situation:

A commercial mini-split system failed to power on, causing significant discomfort during a hot summer period.

Troubleshooting Steps:

1. Electrical Checks: The technician verified that there was power at the electrical panel and that the breaker was in good condition.

2. Continuity Testing: Using a multimeter, continuity tests were performed on the power supply cables to the outdoor unit, which showed a break in the circuit.

3. Component Inspection: Further inspection revealed that a rodent had damaged the wiring inside the outdoor unit.

Solution:

The damaged wires were replaced, and all connections were secured with additional protective conduit to prevent future occurrences. The system was then tested for functionality and started up without issues.

Case Study 3: Frequent Cycling and Increased Energy Bills

Situation:

Homeowners reported that their mini-split system was cycling more frequently than usual and noticed a significant increase in their energy bills.

Troubleshooting Steps:

1. Thermostat and Sensor Checks: Initial checks on the thermostat settings and sensors did not reveal any issues.

2. Inspection of the Outdoor Unit: The outdoor unit inspection showed a heavily soiled condenser coil, which was reducing the system's ability to expel heat effectively.

3. System Performance Tests: Performance tests indicated that the compressor was overworking to try and maintain the set temperature.

Solution:

The condenser coils were thoroughly cleaned, and the surrounding area was cleared of any debris that could restrict airflow. This restoration of proper heat dissipation efficiency allowed the system to return to normal cycling patterns, and subsequent utility bills reflected a more typical energy usage.

Learning Outcomes:

These case studies underscore the importance of regular maintenance, such as cleaning filters and coils, and the need for vigilant observation of system performance indicators like unexpected noises, cycling patterns, and energy consumption.

2. Discussion On Maintaining System Longevity And Performance

Maintaining the longevity and performance of inverter mini-split systems is crucial for ensuring efficient operation, minimizing repair costs, and extending the lifespan of the equipment. This section discusses best practices and key strategies to achieve these goals, drawing from real-world experiences and technical insights.

1. Regular Maintenance Schedule:

   - **Importance**: Adhering to a regular maintenance schedule is paramount. Regular checks and cleaning prevent the buildup of dirt and debris which can impair system efficiency and cause premature wear.

   - **Actions**: Schedule monthly, seasonal, and annual maintenance tasks such as cleaning filters, checking refrigerant levels, inspecting electrical connections, and ensuring that all mechanical parts are in good working order.

2. Proper Installation and Sizing:

- **Impact**: Proper installation and correct sizing of the system are foundational to its efficiency and longevity. An improperly installed or incorrectly sized system will work harder than necessary, which can lead to quicker degradation and higher operational costs.

- **Recommendations**: Always work with qualified professionals for installation and insist on detailed load calculations to ensure the system is perfectly sized for the space.

3. Optimal Use of System Features:

- **Energy Efficiency**: Utilize the advanced features of inverter mini-splits, such as programmable timers, eco modes, and temperature zoning. These features are designed to optimize energy use and maintain comfort without overburdening the system.

- **Practical Tips**: Educate users on how to use these features effectively. For instance, setting the thermostat to a reasonable temperature, using timers to reduce operation when not needed, and adjusting zones based on occupancy.

4. Addressing Issues Promptly:

- **Early Detection**: Encourage the early detection of issues such as unusual noises, leaks, or performance dips. The sooner a problem is identified, the less likely it is to cause significant damage.

- **Proactive Repairs**: Make repairs promptly to prevent minor issues from developing into major problems. Use qualified service technicians to ensure repairs are performed correctly.

5. Environmental Considerations:

- **Location**: Protect outdoor units from extreme weather conditions and ensure they are placed in well-ventilated areas free from potential blockages like leaves, snow, or debris.

- **Climate Control**: In regions with extreme temperatures, additional measures like installing protective covers or using climate-adapted lubricants can help maintain system efficiency.

6. Technological Upgrades:

- **Up-to-Date Technology**: Keep abreast of technological advances in HVAC systems, including new refrigerant types that are more environmentally friendly and systems that offer better performance metrics.

- **Retrofitting**: Consider retrofitting older systems with new technologies to enhance their efficiency and extend their useful life.

Maintaining the longevity and performance of inverter mini-split systems is a multifaceted endeavor that involves a combination of proper installation, regular maintenance, and user education. By implementing these best practices, stakeholders can ensure that their systems operate efficiently over their expected lifespan, providing reliable comfort and minimizing environmental impact. These strategies not only protect the investment made in HVAC equipment but also contribute to a sustainable approach to heating and cooling needs.

Chapter 9: The Future Of Inverter Mini-Splits

1. Innovations In Inverter Technology And Mini-Split Systems

The HVAC industry is continually evolving, with technological advancements and environmental considerations driving significant changes in the design and functionality of inverter mini-split systems. Here's a look at the future directions and innovations in this technology:

1. Increased Energy Efficiency:

  - **Advanced Inverter Technology**: Future developments are expected to enhance the efficiency of inverter technology even further, reducing energy consumption and increasing cost-effectiveness. Innovations may include more precise control algorithms and better variable-speed compressor designs that optimize energy use throughout the operation.

- Higher SEER Ratings: As regulatory requirements tighten, the Seasonal Energy Efficiency Ratio (SEER) of mini-split systems will continue to rise, leading to even greater energy savings and lower environmental impact.

2. Integration with Smart Home Technologies:

   - IoT Connectivity: Inverter mini-splits are increasingly becoming integrated with the Internet of Things (IoT), allowing for smarter, more responsive climate control systems. These systems can learn from user behaviors, adapt to real-time weather conditions, and be controlled remotely via smartphones or voice-activated home assistants.

   - Predictive Maintenance: Future systems could incorporate predictive analytics to forecast when maintenance is needed or when components are likely to fail, thus preventing downtime and extending system life.

3. Enhanced Indoor Air Quality Features:

 - **Air Purification Technologies**: Newer models are expected to incorporate advanced air purification technologies, such as HEPA filters, UV lights, and ionizers, directly into mini-split systems. This integration will provide not only temperature control but also ensure cleaner indoor air, which is particularly important in areas with high pollution levels or for allergy sufferers.

 - **Humidity Control**: Improved humidity control capabilities will be a focus, with systems able to finely control the moisture levels in the air, enhancing comfort and preventing mold growth.

4. Eco-Friendly Refrigerants:

 - **Low GWP Refrigerants**: As environmental regulations become stricter, the shift towards refrigerants with low Global Warming Potential (GWP) will accelerate. These refrigerants are not only better for the environment but are also designed to be more energy-efficient.

- **Leak Detection and Prevention**: Advanced leak detection systems will become standard, reducing the risks associated with refrigerant leaks, both from an environmental and system efficiency perspective.

5. Modular and Scalable Designs:

- **Customizable Systems**: Future mini-split systems will offer greater modularity, allowing systems to be more precisely tailored to the specific needs of a home or building. This customization will enable optimal efficiency by matching the exact heating and cooling needs without overcapacity.

- **Expandable Systems**: Technologies that allow systems to be easily expanded, such as adding more indoor units without needing significant modifications to existing installations, will become more prevalent.

6. Aesthetic Integrations:

- **Design Innovations**: As mini-splits become more common in a variety of settings, including high-end homes and businesses, the demand for systems that can blend seamlessly with interior designs will increase. Expect to see a range of customizable designs, from minimalistic and hidden units to those that serve as aesthetic features in their own right.

The future of inverter mini-split systems is bright, with ongoing innovations aimed at improving efficiency, connectivity, environmental friendliness, and user-friendliness. These advancements promise to make inverter mini-splits an even more attractive option for both residential and commercial properties, ensuring that they play a crucial role in the future of sustainable climate control solutions.

2. Environmental Impact And Energy Efficiency Trends

The environmental impact of heating, ventilation, and air conditioning (HVAC) systems, particularly inverter mini-split systems, is a growing concern that is driving significant changes in technology and regulatory standards. Here's an overview of the trends in energy efficiency and environmental stewardship that are shaping the future of these systems:

1. Increasing Energy Efficiency Standards:

   - **Higher SEER Ratings**: There is a clear trend towards requiring higher Seasonal Energy Efficiency Ratio (SEER) ratings for HVAC systems. These improvements in energy efficiency not only reduce electricity consumption but also lower the carbon footprint of residential and commercial buildings.

- **Minimum Efficiency Reporting Value (MERV) Ratings**: Improvements in filtration technology are expected to increase, with a focus on reducing particulate emissions and improving indoor air quality, further contributing to environmental health.

2. Reduction in Greenhouse Gas Emissions:

- **Low Global Warming Potential (GWP) Refrigerants**: As part of international efforts to combat climate change, there is a move towards refrigerants with lower GWP. This shift is expected to accelerate, with newer, more environmentally friendly refrigerants becoming the standard.

- **Advanced Leak Detection**: Technologies for detecting refrigerant leaks, which contribute to both ozone depletion and global warming, are becoming more sophisticated and widespread. This helps ensure systems are not only more efficient but also less harmful to the environment.

3. Integration with Renewable Energy Sources:

- **Solar-powered Mini-Splits**: There is an increasing interest in integrating mini-split systems with solar power. This integration allows these systems to operate in a more eco-friendly manner, reducing reliance on grid power and decreasing greenhouse gas emissions.

- **Hybrid Systems**: Future developments may focus on hybrid systems that can switch between renewable energy and traditional energy sources to optimize efficiency and reduce environmental impact.

4. Smart Systems and IoT Integration:

- **Energy Consumption Optimization**: The integration of HVAC systems with smart home technologies and IoT is set to increase. These systems can optimize energy consumption based on usage patterns, weather conditions, and real-time energy pricing.

- **Automated Energy Management**: Advanced algorithms can manage heating and cooling needs more dynamically, reducing energy consumption during off-peak hours or when rooms are unoccupied.

5. Regulatory and Policy Changes:

- **Building Codes and Standards**: Changes in building codes and energy efficiency standards are likely to mandate the use of high-efficiency HVAC systems. These changes will encourage the adoption of the latest inverter technologies that comply with these new standards.

- **Incentives and Rebates**: Governments and energy companies may offer more incentives and rebates for installing energy-efficient systems, which will promote the uptake of high-efficiency mini-splits.

6. Consumer Awareness and Demand:

- **Educated Consumers**: As consumers become more aware of environmental issues, there is a growing demand for HVAC systems that not only provide comfort but also do so sustainably.

- **Market Response**: Manufacturers are responding with systems that are easier to recycle, use fewer hazardous materials, and incorporate eco-friendly technologies.

The trend towards environmentally friendly and energy-efficient HVAC systems is clear and growing. Inverter mini-split systems are at the forefront of this trend, offering solutions that reduce energy consumption and environmental impact. As technology advances, these systems are expected to become even more integrated with smart technologies and renewable energy sources, leading the way in sustainable climate control solutions.

Appendix

Glossary Of Terms

This glossary provides definitions for terms commonly used in relation to inverter mini-split systems, aiming to enhance understanding of the technical language used throughout the book.

1. Inverter Technology: A technology that allows the compressor to operate at variable speeds by adjusting the power supply frequency, enabling it to continuously regulate the temperature by varying the motor speed without turning off.

2. Mini-Split System: A type of ductless heating and cooling system that consists of an outdoor compressor/condenser unit and one or more indoor air handling units, connected by refrigerant lines and electrical wiring.

3. SEER (Seasonal Energy Efficiency Ratio): A measure of the cooling efficiency of an air conditioner over an entire cooling season. It is calculated by dividing the cooling output by the total electrical energy input during the same period.

4. HSPF (Heating Seasonal Performance Factor): A measure of the heating efficiency of heat pumps over an entire heating season. It is calculated similarly to SEER but for heating.

5. Refrigerant: A substance used in a heat cycle to transfer heat from one area to another. It is capable of vaporization and condensation within the operating temperatures of the refrigeration system.

6. Compressor: A component in the mini-split system that compresses the refrigerant, raising its temperature and pressure, and moves it through the system.

7. Condenser: A component of the mini-split system located in the outdoor unit where the refrigerant releases the heat it absorbed from the indoor air.

8. Evaporator: Located inside the indoor unit, it absorbs heat from the indoor air, allowing the refrigerant to evaporate and cool the surrounding area.

9. Refrigerant Lines: Tubes or pipes that carry refrigerant between the indoor and outdoor units of a mini-split system.

10. Air Handler: The indoor component of a mini-split system that includes a coil and a fan. The fan blows indoor air across the coil to facilitate heat exchange.

11. Remote Control: A device used to control the settings of a mini-split system, including but not limited to temperature, fan speed, and operating mode.

12. GWP (Global Warming Potential): A measure of how much heat a greenhouse gas traps in the atmosphere compared to carbon dioxide

(CO_2). Used to represent the potential climate impact of refrigerants.

13. IoT (Internet of Things): The network of physical objects—"things"—that are embedded with sensors, software, and other technologies for the purpose of connecting and exchanging data with other devices and systems over the internet.

14. Ductless System: A heating and cooling system that does not require a network of ducts to distribute air. Mini-splits are a common type of ductless system.

15. Heat Pump: A device used for either heating or cooling a space by moving heat between two reservoirs. In the context of mini-splits, it refers to systems that can both cool and heat indoor spaces.

Frequently Asked Questions (FAQs)

The following are common questions about inverter mini-split systems, providing quick and clear answers to help users understand and effectively use their HVAC systems.

1. What is an inverter mini-split system?

- An inverter mini-split system is a type of ductless air conditioning system that uses an inverter-driven compressor to continuously adjust the cooling or heating output to the needs of the space, enhancing energy efficiency and comfort.

2. How does an inverter mini-split differ from a traditional air conditioner?

- Unlike traditional air conditioners that operate with compressors turning on and off at full capacity, inverter mini-splits have compressors that vary their speed to maintain consistent temperature levels, reducing energy consumption and wear on the system.

3. Can an inverter mini-split heat and cool?

- Yes, many mini-split systems are designed as heat pumps, which means they can both heat and cool a space depending on the settings.

4. What are the benefits of installing a mini-split system?

- Mini-split systems offer several benefits including lower energy costs, easy installation without the need for ductwork, individual zoning capabilities, and reduced noise levels compared to traditional HVAC systems.

5. How often does a mini-split system need maintenance?

- It is recommended to clean the air filters monthly or as needed, perform a detailed check of the entire system annually, and schedule professional maintenance at least once a year to ensure optimal performance.

6. Can I install a mini-split system myself?

- While it's technically possible to install a mini-split system as a DIY project, professional installation is strongly recommended to ensure the unit operates efficiently, complies with local building codes, and qualifies for warranties and energy rebates.

7. How long do mini-split systems typically last?

- With proper maintenance, mini-split systems can last 15 to 20 years, depending on usage patterns and environmental conditions.

8. Are mini-split systems environmentally friendly?

- Yes, mini-split systems are considered environmentally friendly due to their high energy efficiency and the use of newer refrigerants with lower global warming potentials (GWP).

9. What should I do if my mini-split system isn't cooling or heating properly?

- Check if the filters are clean, ensure there are no obstructions to airflow, and verify that the system settings are correct. If problems persist, contact a professional technician to check for issues such as refrigerant leaks or compressor problems.

10. How much does it cost to install a mini-split system?

- Installation costs can vary widely based on factors such as the number of indoor units, the specific model chosen, and the complexity of the installation. Generally, costs can range from $1,500 to $8,000 or more.

11. What size mini-split system do I need?

- The size of the system you need depends on various factors including the size of the space, insulation quality, climate, and specific heating and cooling needs. It's best to have a professional perform a detailed load calculation to determine the appropriate size.

12. Can mini-splits be used in cold climates?

 - Yes, many modern mini-split systems are equipped with technology that allows them to operate efficiently in cold climates, but it's important to choose a model specifically designed for low-temperature operation.

www.ingramcontent.com/pod-product-compliance
Lightning Source LLC
Chambersburg PA
CBHW050103230526
45470CB00004B/1661